Copyright ©2016 Debbie Lyn
All Rights Reserved
First Printing
ISBN 9780692713174

Made in the USA
Monee, IL
08 November 2026

67377455R00090